PAIN MANAGEMENT IN AYURVEDA

By-

Dr. MANISHA RANI

B.Sc. Hons. (Chemistry), B.A.M.S. (Muzaffarpur)

Dr. ANKUR SINHA

B.A.M.S, M.D. (PUNE), D.Y.A (Yoga Advanced),
E.M.S., Ex. Asst. Prof. S.D.A.Med. College & Hospital,
Ranchi

Dr. ANIL KUMAR

M.D. (Ay.), B.H.U., Associate Prof. Dept. of Roganidan,
Saraswati Ay. College & Hosp., Mohali

Table of contents

CHAPTER 1

Introduction

Everybody must have experienced some kind of pain in their life. Pain as such can be described according to "The International Association for the Study of Pain" as "It is an unpleasant sensory and emotional experience associated with actual or potential tissue damage, or described in terms of such damage".

Pain is something that can only be experienced by the person affected by pain. For total relief from pain, the doctor must know the intensity of pain from which the patient is suffering. Doctor must ask about the nature, severity, timing, location, quality, aggravating & relieving factors of the pain. Based on the above factors, medication can be administered properly. So as to understand the intensity and proper medicines that should be administered, pain must be scaled. Considering all these facts, it may be classified according to the pain assessment scale.

Pain intensity	Word scale	Nonverbal behaviours
0	No pain	Relaxed, calm expression
1-2	Least pain	Stressed, tense expression
3-4	Mild pain	Guarded movement,

		grimacing.
5-6	Moderate pain	Moaning, restless
7-8	Severe pain	Crying out
9-10	Excruciating pain	Increased intensity of above.

Ayurveda says, pain may be either physical or mental. Mental pain can be cured by counselling & medications. Physical pain can be cured with medications & other therapies.

It is seen in today's practice, patients are very much addicted to NSAIDs (Non Steroidal Anti Inflammatory Drugs) due to pain they suffer. There must be some indications where NSAID's should be prescribed.

NSAIDs are generally used for the symptomatic relief of the following conditions:

- Osteoarthritis.
- Rheumatoid arthritis.
- Mild-to-moderate pain due to inflammation and tissue injury.
- Low back pain.
- Inflammatory arthropathies (e.g., ankylosing spondylitis, psoriatic arthritis, reactive arthritis)
- Tennis elbow.
- Headache.
- Migraine.

- Acute gout
- Dysmenorrhoea (menstrual pain)
- Metastatic bone pain
- Postoperative pain
- Muscle stiffness and pain due to Parkinson's disease
- Pyrexia (fever)
- Ileus
- Renal colic
- They are also given to neonate infants whose ductus arteriosus is not closed within 24 hours of birth.
- And many more other conditions

NSAIDs & Ayurvedic drugs are usually indicated for the treatment of acute or chronic conditions where pain alone or with inflammation is present.

Mechanism of action of NSAIDs.

Most NSAIDs act as nonselective inhibitors of the enzyme cyclo-oxygenase (COX), inhibiting both the cyclo-oxygenase-1 (COX-1) and cyclo-oxygenase-2 (COX-2) isoenzymes. This inhibition is competitively reversible, compared to the mechanism of action of aspirin, which is irreversible inhibition. Prostaglandins act as messenger molecules in the process of inflammation.

An enzyme called COX-1 is having a great role in regulating many normal physiological processes in the body. One of these is in the stomach lining, where prostaglandins prevent the stomach mucosa from being eroded by its own acid, thus having a protective role. It is the inhibition of COX-2 that produces the desirable effects of NSAIDs.

When nonselective COX-1/COX-2 inhibitors (such as ibuprofen, aspirin etc.) lower stomach prostaglandin levels, then ulcers of the stomach or duodenum & internal bleeding can result.

The discovery of COX-2 led to the research for the development of selective COX-2 inhibiting drugs that do not cause gastric problems.

Paracetamol (acetaminophen) is not considered as NSAID because it has little anti-inflammatory activity. It treats pain mainly by blocking COX-2 mostly in the central nervous system.

Still, there are many unexplained mechanism of action of NSAIDs which are under study now.

Till now we have read many good uses of NSAIDs. Though the NSAID's have tremendous good effects for above mentioned conditions. But every substance has some pros and cones. Same way NSAID's are also having some sort of side-effects. Most of the patients are aware of the side-effects that arise from the NSAID's. But still they take these drugs so as to get relief from the pain as early as possible.

As said above, NSAID's have some sort of side effects. So, what are these side-effects?

Common side effects of NSAIDs:

- Stomach ulcers and gastrointestinal bleeding
- Increased blood pressure
- Delayed digestion
- Dizziness
- Tinnitus (ringing in the ear)
- Headache
- Depression
- Kidney damage
- Erectile dysfunction

After knowing so many side-effects, still people continue to take these drugs. So, "is there any alternative treatment which can help the patients to get relief from pain without any side effects?"

Yes, definitely. Ayurveda has the answer for this. Without internal medications also, some Ayurvedic therapies undeniably cures pain which can give relief by 60-100%. Such alternative therapies are actually not alternative. Actually these are one of the best treatments considered in Ayurveda. Such therapies are –

- Agnikarma Therapy
- Viddha Karma Therapy
- Marma Therapy
- Other localised treatments (*Kati Basti, Griva Basti...*)

For the wellbeing of humans, such successful therapies should be practiced regularly. In this book, I have taken efforts to keep forward the concepts of these therapies as much as I can.

Before proceeding further, base line ideas about these Karmas (procedures) must be understood.

Agnikarma Therapy

Agni means fire

Karma means procedure

So, **Agnikarma** means treatment procedure where fire is being used. The modern development of this age old concept is "Cauterisation" which is now the main equipment during surgery. Various painful conditions like joint pain, low backache, head ache, schizophrenia & psychosomatic disorders, abdominal cramps / discomfort, frozen shoulder & few convulsive disorders like epilepsy etc. can be treated with this therapy.

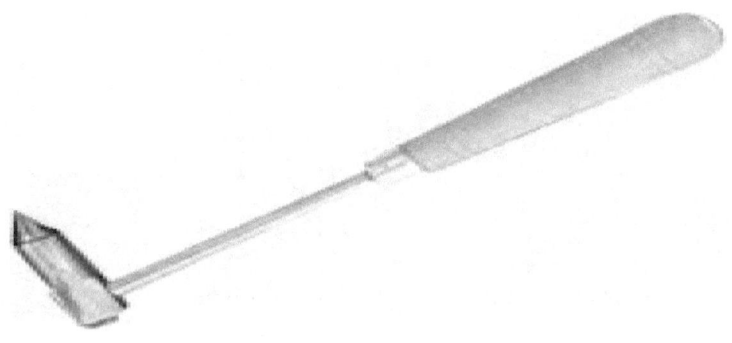

Figure 1: Agnikarma Shalaka (Instrument for doing Agnikarma)

Viddha (Vedhana) Karma Therapy

Viddha Karma is a kind of blood-letting therapy for the management of pain. This is done by the help of a needle (e.g. scalp vein, etc.) that is to be inserted at the affected site.'

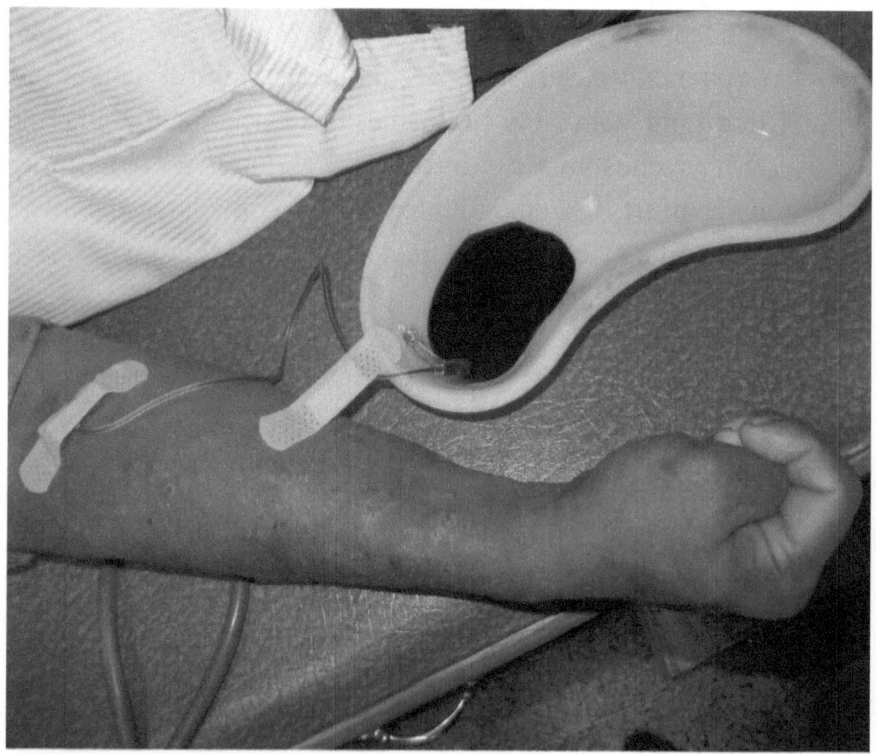

Figure 2: Viddha Karma (Blood letting therapy)

Marma Therapy

Marmas are specific points on the body where the application of pressure induces the flow of vital energy (Prana) along a complex system of subtle channels called (Nadis).

Kati Basti, Griva Basti, etc...

Some other specific local therapies are also mentioned in Ayurveda which helps a lot in pain management. For example; pain of Cervical Spondylosis, Chronic pain in the neck region, sciatica, slipped and degenerative discs, back pain etc. More details will be given in the respective chapters.

Similar to above mentioned therapies there are many more therapies to overcome physical pain.

There are certain other therapies which help in combating mental pain too. For e.g. Shirodhara.

CHAPTER 2

Role Of Viddha Karma In The Management Of Pain.

Introduction:

The surgery based text "Sushruta Samhita" has a separate chapter on "Siravyadha Vidhi Sharira" in Sharira Sthana. After going through this chapter, many practical concepts have been explored. Siravedha Vidhi is the process of doing venipuncture.

There are many conditions where the patients suffer from pain caused by vitiated blood. These conditions can be cured if the vitiated blood is removed from the body. One such method is Siravedha.

Sushruta Samhita has given a beautiful verse (Su. Sha. 8/22) which explains "the diseases get cured very rapidly by Siravyadha as compared to other therapies like oleation therapy, anointments etc."

One more verse taken from the Sushruta Samhita Sharira Sthana 8/23 states "As Basti is considered as half treatment of Kaya Chikitsa (Ayurvedic medicine), the same way Siravyadha is considered as a half treatment of

Shalya Chikitsa (Surgery)." This verse shows the importance of Siravyadha in practice.

But before going in details about the practical approach of Viddha Karma, one must know some basic ideas regarding its contra-indications, precautions that is to be taken, regimens that should be followed before & after the procedures etc.

Contra-indications for Siravedha (Su. Sha 8/3)

Siravyadha should not be done in the patients listed below -

- Child (very young)
- Very old
- Having dryness in the body
- Injured
- Emaciated person
- Fearful person
- Tired
- Indulged too much in sex
- Intoxicated
- After excessive travelling
- Patients underwent Vamana, Virechana & Asthapana
- Not taken sound sleep
- Impotent
- Thin person
- Pregnant lady
- Patient suffering from Kaasa, Swasa, Shosa, Pravridha Jvara, Akshepaka, Pakshaghata, Upvasa, Trishna, Murchita.

Venipuncture should also be avoided when the weather is too cold, too hot, windy or cloudy. Venipuncture should not be done even in healthy. (Su. Sha. 8/7).

In case of person who are fainted, scared, exhausted, thirsty & also when the veins are not raised & stabilised, blood does not flow out of the veins. In case of fainting etc, Vata blocks the openings of the blood vesselss. Hence, the proper discharge of blood does not take place. Therefore, venipuncture should also be avoided in these cases. (Su. Sha. 8/13)

Other contra-indications for Siravaidha (Su. Sha. 8/3)

Venipuncture should not be done in child, old aged people, who are wasted due to wounds in chest, fearful, tired, alcoholic, emaciated due to excessive walking & sex; individuals who went for Vamana (emesis), Virechana (purgation) & unctuous enema; individuals who are awake, impotent, emaciated, pregnant & those suffering from Kasa (cough), Swasa (dyspnoea), Shosa (consumption), Pravriddha Jvara (hyper-pyrexia), Akshepa (convulsions), Pakshaghata (hemiplagia), Upvasa (having fast), Pipasa (thirst), Murcha (fainting). Not only this, veins which could cause complications or death on puncture should also be avoided from being punctured. Furthermore, veins which are puncturable but invisible, visible but can't be stabilised or though can be stabilised, but can't be raised should not be punctured.

Regimen to be followed before the procedure (Su. Sha. 8/6)

Patient should be posted for Snehana & Svedana first. Then the liquid diets or Yavagu (Gruels) should be prescribed which should be prepared from the drugs having inhibiting actions on the vitiated Doshas. After this regimen, patient should be posted for Siravedha.

Indications for Sira-Vyadha

Venipuncture should be done in the diseases which can be cured by blood-letting. (Su. Sha. 8/4).

Even in cases where venipuncture is contraindicated, it is allowed in cases of emergency conditions & poisoning. (Su. Sha. 8/5).

Diseases are cured by Venipuncture faster than other procedures like Snehana (unction) & Lepas (paste) etc. (Su. Sha 8/22).

In conditions, where Dosas are remarkable & patient's condition is weak or fainted, should be again posted for venipuncture in afternoon, next day or on the 3rd day. (Su. Sha. 8/14).

Quantity of blood to be taken out–

1 Prastha = 13 ½ Pala = 540ml. *(Su. Sha. 8/16)*

Ideally 1 Prastha = 16 Palas, but for the process of blood-letting, 1 Prastha is considered as 13 ½ Pala.

Practically, in today's practice I used to take out 125 – 150 ml. This depends upon the body strength of the patients.

Requirements for Viddha Karma

- *Viddha Karma*
  1. O.P.D. Room or patient's house
  2. Insulin Needle (No. 26)
  3. Scalp Needle (No. 18-20)
  4. Measuring jar
  5. B.P.Apparatus

Venipuncture is considered as half of the treatment in surgery, the same way as Basti is considered as half of the treatment in general medicine. (Su. Sha. 8/23).

Instruments for Viddha Karma

Horn, gourd, leech & scarification (Prakshana) in retrograde order depending upon the situation should be used for venesection in case of vitiated blood. **(Su. Sha. 8/25)**.

Specific indications for these instruments (Su. Sha. 8/26)

- In deep seated conditions – Leech is favoured.
- In localised lump – Scarification (Prakshana) is helpful.
- In generalised blood impurity – Venipuncture is preferred.
- In diseases of skin – horn & gourd are used.

Imperfect puncturing (Su. Sha. 8/18)

Imperfect puncturing is of twenty types.

- **Deficient puncturing** – if the vein is punctured with small instrument than required, proper blood flow does not takes place & this also leads to pain & swelling.
- **Excessive puncturing** – there is excessive outflow of blood.
- **Crooked**
- **Crushed** – if the veins are punctured again & again with blunt instruments, the veins get crushed.
- **Lacerated** – the veins get lacerated if invisible vein is punctured repeatedly on both ends.
- **Non-discharging** – no blood flow due to cold, fear or fainting.
- **Excessive discharging** – if too sharp & big tipped instruments are used for puncturing the veins, severe discharge of blood takes place.

- **Marginal puncture** – leads to less discharge of blood.
- **Dried (Parisushka)** – the veins which gets filled with air on puncturing are dried. This happens due to scarcity of blood in the body as in case of low level of blood in body.
- **Insufficient puncture** – occurs when only one fourth portion is punctured, leading to scanty blood flow.
- **Shaking** – discharge of little amount of blood due to ligature at improper place of unstable veins.
- **Unraised when punctured** – same thing happens as in shaking, if unraised veins are punctured.
- **Injured by instrument** – due to injury to the veins, excessive bleeding takes place.
- **Obliquely punctured** – when veins are punctured obliquely, it is said to be obliquely punctured.
- **Unpunctured** – by use of low-grade instruments, veins are not properly punctured (unpunctured) & gets injured.
- **Non puncturable Siras** – are those veins, which should not be punctured, otherwise it may lead to deformities of death.

- **Punctured while unsteady** – veins, if punctured while it is unsteady may lead to other complications.
- **Frequently discharging** – vein from which blood flows out frequently by excessive striking of the spot.
- **Frequently punctured** – is one, which is punctured frequently by small instrument.
- **Punctured at vascular ligament, bones, joints & other Marmas** causes pain, swelling, disability & death.

Methods for ligation for performing venipuncture (Su. Sha. 8/8)

Applying tourniquet wherever required is a must. This helps the veins to be gorged near the site of ligation & thus makes it easy to be visualised & punctured.

Depth of Vyadha karma (Su. Sha. 8/9)

1. Vyadha in Mamsa should be as deep as 1 Yava (Barley) with Vrihimukha instrument.
2. Vyadha in bones should be as deep as ½ Yava with Kutharika instrument.

But practically I use Insulin needle No.26, having size .45mm (breadth) & 13mm (length). Piercing is done   1. For skin – 2 to 4 mm deep.

2. For Mamsa – 4 to 6 mm deep.

3. For Snayu, Asthi, Sandhi – 6 to 10 mm deep.

Marma consideration

Before practicing the Viddha treatment, one must have a fine knowledge of all the Marmas in the body, specially its exact location. Injuries to the Marma cause many complications, in extreme cases it also leads to death of the patient.

To reduce these complications we started these techniques by Insulin needle No. 26 with length 13 mm. This needle cuts the chances of getting complications many times.

Postures for Vyadha Karma (Su. Sha. 8/8)

1. Veins in hand- It should be punctured by raising the hand above with fist closed having thumbed inside & ligation should be done above the point of Viddha Karma.
2. In case of Grdhrasi (Sciatica), vein is punctured while patient flexes his knees.
3. In case of Visvaci (Brachial Neuritis) the vein is punctured while patient flexes his elbow.

4. For venipuncture in hip, back & shoulder, the patient should sit with neck flexed & back extended, and then the vein should be punctured.

5. In penis, veins should be punctured while erect penis is bent.

6. Veins in the lower surface of the tongue should be punctured while raising the tongue & keeping it fixed by the help of upper incisors teeth.

7. Mouth should be kept wide open while puncturing palate & gums.

8. For venipuncture in abdomen & thorax, body should be extended, chest should be expanded & head rose so as to increase the muscle tone.

9. In case of venipuncture of palate & gums, mouth should be opened widely.

10. The leg that has to be punctured should be tied below the knee joint while keeping ankle joint pressed with hands & applying tourniquet four fingers above the point where the vein have to be punctured.

Actual points for the Viddha Karma (Su. Sha. 8/17)

In case of

- **Pada-daha**

- **Pada-harsa**
- **Cippa**
- **Visarpa** (Erysipelas)
- **Vata-Rakta** (Gout)
- **Vata-Kantaka**
- **Vicarchika** (Eczema)
- **Pada-dari** (Cracked foot)
- Etc...

Viddha should be carried out about two fingers above the Kshipra Marma.

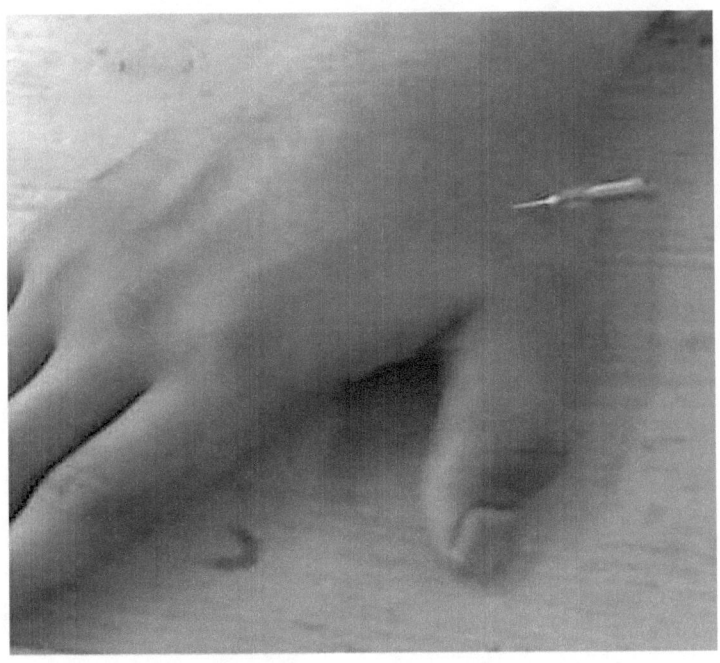

2 cm above Kshipra Marma

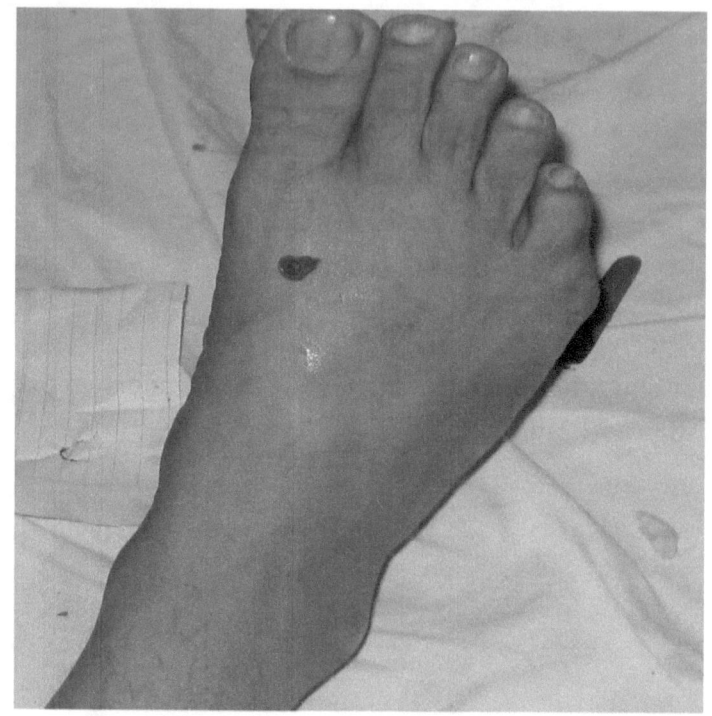

2 cm above Kshipra Marma

Kshipra Marma

Location :

It is situated between the big toe & first toe.

Measures: ½ Anguli

Type of Marma:

Kalantara Pranahara Marma (i.e. death after some time).

Signs of injury:

impairment in functions of flexion & adduction of the great toe.

Severe bleeding may occur if there is damage to the artery. Further this may result in unconscious, low B.P. & death.

Vascular: Dorsal venous network of the foot.

Dorsal Metatarsal artery.

Nervous: Branches of deep peroneal Nerve.

In case of **Shlipada** (Elephantiasis), Su. Chi. 19/52-56

- **Vataja Shlipada** – Sira four fingers above the ankle joint should be punctured.
- **Pittaja Shlipada** – Sira four fingers below the ankle joint should be punctured.
- **Kaphaja Shlipada** – Sira present on the dorsum of greater toe, in between the first

inter-phalangeal joint & the proximal border of the nail should be punctured.

Anatomy:

1) From the dorsal venous arch, the great saphenous vein passes anterior to the medial malleolus of the ankle and enters the medial side of the leg
2) Nerves – Medial crural cutaneous nerve & tibial nerve on its deeper side.

On the dorsum of the great toe

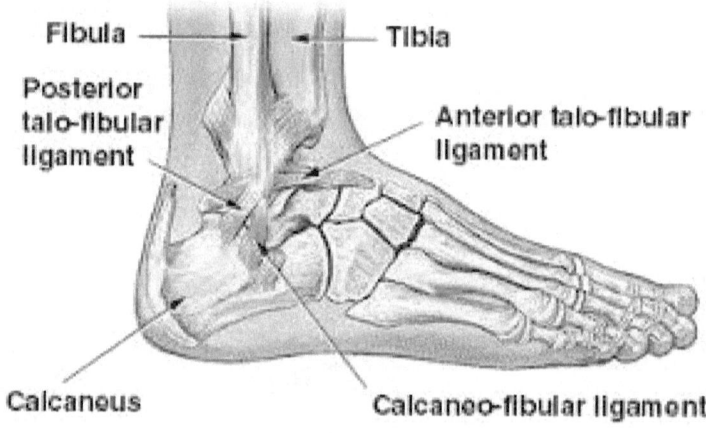

In case of

- **Krostukasiras**

- **Khanja**
- **Pangu**
- **Other related Vatika diseases**
  Viddha should be carried at the point which should be four fingers above the ankle joint.

In case of **Apachi** (Scrofula), Viddha Karma should be done two fingers below the Indrabasti Marma both in forearms & legs.

Indications - Apachi (Scrofula) – Neck & Axillary.

- Inguinal

Anatomy for Apachi:

Two fingers below the Indrabasti Marma of the foreleg –

Vascular: Branches of Posterior Tibial Artery & Veins,

Small Sephanous Vein.

Nervous: Superficial Peroneal Nerve

Two fingers below the Indrabasti Marma of the forearm –

Vascular: Mascular branches of Radial Artery & Cephalic Vein.

Indrabasti Marma:

Location:

At the centre of the line joining Posterior surface of Calceneum & centre of Popliteal fossa.

Measure: ½ Anguli

Type of Marma: Kalantara Pranahara Marma.

Signs of injury:

Impairment in the function of the foot.

Severe bleeding occurs if the artery at this point is damaged. This may lead to collapse, low B.P. & Shock.

In case of

- **Grudhrasi** (Sciatica)
- **Vishvachi** (Brachial Neuritis)

Viddha Karma should be performed at two sites; i.e. four fingers above & below the knee & elbow joints respectively.

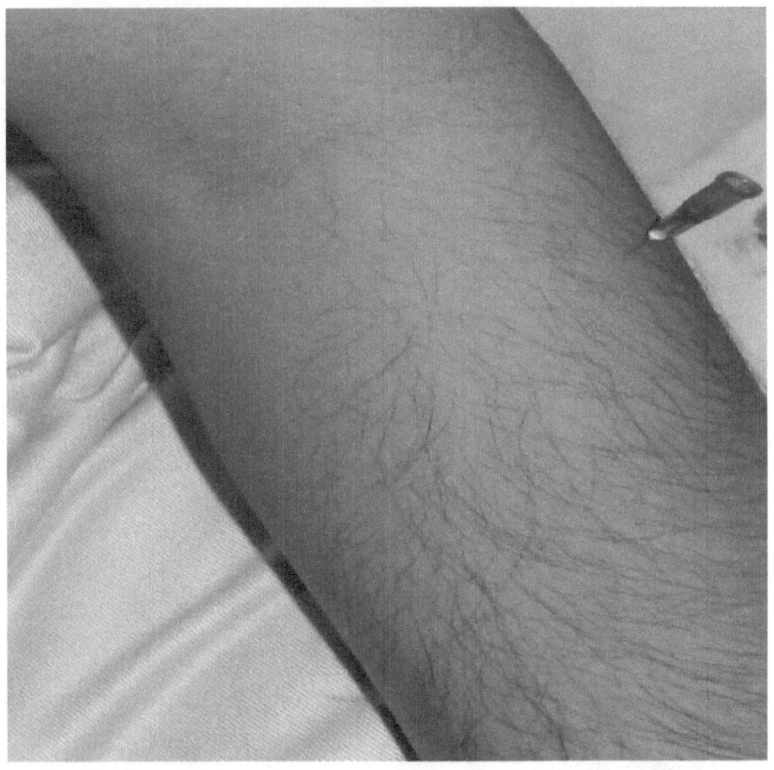

4 fingers below elbow joint

In case of **Galaganda** (Goitre), the vein situated at the root of thigh & root of Axilla should be punctured. It should be performed on both the sides.

In case of **Pleha-odara** (Spleenomegaly), vein should be punctured in the left arm, at the middle & inner aspect of the elbow joint (practically below or above the joint) or between little & index fingers.

Karpura Marma:

Location: Elbow

Type of Marma: Vaikalyakara Marma

Measure: ½ Anguli.

Signs of injury:

- Impairment of functions of the forearm
- If artery is ruptured, may cause severe haemorrhage.
- Low B.P.

In case of

- **Yakrutodara** (Hepatomegaly).
- **Kaphodara**
  Same procedure on the right side should be followed as mentioned in Plehodara.

In case of

- **Kasa** (Cough)
- **Swasa** (Dyspnoea)
  Same point as mentioned above should be selected for puncturing.

In case of **Pravahika** (Dysentery with pain), Viddha Karma should be carried out two fingers medial to Anterior Superior Iliac spines.

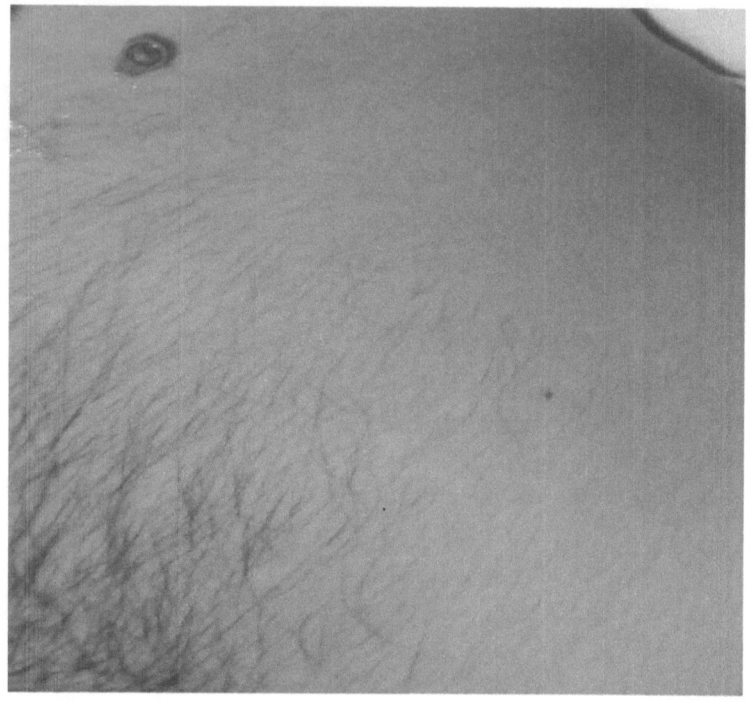

Fig: Two fingers medial to Anterior Superior Iliac spines.

Anatomy for this region:

Vascular: 10th & 11th intercostal artery & vein

Nervous: 10th & 11th intercostals nerve.

In case of

- **Parivartika** (Paraphimosis)
- **Updamsa** (Soft chancre)l
- **Shukadosa**
- **Shukra-Vyapata** (Defects in the Sperms or Semen).

Viddha Karma should be performed in the middle of the penis by keeping the penis dorsiflexed (Dorsiflexion means upward or backward flexion of a part of the body)

In Updamsa, it should be carried out after the unction & sudation is done by the help of leeches, if it is superficial. In case of deep lesions, Sira Moksha should be done. (Su. Chi. 19/25)

In Paraphimosis, always there is oedema around corona of Glans penis due to venous obstruction. It is a clear fluid. In practice, numerous punctures are performed to remove the fluid which helps in the reduction of prepucial skin.

In case of **Mutra-vriddhi** (Hydrocele), Viddha Karma should be carried out on the lateral side of scrotum. The fluid in case of hydrocele should be drained by puncturing the scrotum.

In case of **Dakodara** (Ascites), puncture should be done four fingers below the umbilicus & four fingers away from the raphe in the left side. Puncture is to be done to remove the fluid (tapping). All the fluid should not be removed at a single sitting due to fear of rupture of blood vessels as Sroto-riktata will result in accumulation of Vata.

Nabhi (umbilicus) Marma:

Measure: 4 Anguli

Type of Marma:
Sadyha Pranahara Marma.

Location:
Behind & around the umbilicus

Vascular: Inferior Epigastric Artery & Vein.
Abdominal Aorta.
Inferior Vena cava.

Sign of injury:
Since it is a Sadhya Pranahara Marma, death is sudden. Superficial injury doesn't cause death. If the Abdominal aorta is ruptured, then it may cause instant death due to haemorrhage (bleeding), low B.P. & shock.

In case of-

Antha-Vidradhi (internal abscess or lung abscess) & pleural pain in the sides, puncture between the mid-axillary line & lateral border of breast of the effected side should be performed.

U.S.G. help may be taken for proper guidance.

This treatment should be performed for the drainage of pus only when the abscess is in Pakva-avastha (fully matured).

In case of Aama-Avastha (initial im-mature phase) of the abscess, vein below the elbow joint be cut & blood-letting should be carried out on the sides depending upon the side of pathology (pleural pain & abscess).

In cases of

- **Bahusosa** (Muscular wasting of shoulders)
- **Avabahuka** (Frozen shoulder)

venipuncture between the anterior & posterior borders of the shoulder joint should be done .

In case of-

Triteeyaka Jvara (Terstian fever) should be treated by puncturing at the centre of Trika Sandhi. The point is below the spinous process of fifth lumbar vertebra.

Chaturthaka Jvara (Quartan fever) should be treated by puncturing at mid-axillary line. It is situated below the shoulder joint.

In **Apasmara** (Epilepsy), puncture should be performed near Temporo-Mandibular joint (precaution should be taken, not to injure Temporo-Mandibular joint as it is a Marma point) at the junction of hair line & Shankha Pradesh.

In case of **Unmada** (insanity), puncture of the vein situated in the junction of temples & hair-line & also in Ura (chest), Apanga (outer corner of the eye) & Lalata (forehead) should be done.

Apanga Marma:

Location:
Depression of the lateral end of eyebrow(the outer corner of the eye).

Measure: ½ Anguli.

Type of Marma: Vaikalyakara Marma.

Vascular: Frontal branches of the superficial temporal artery & vein.

Signs if injured:
Injury may lead to blindness & deformities of the face.

Avarta (Lalata) Marma:

Location:

- One finger above the midpoint of the eyebrow on the forehead.
- In the upper border of orbital cavity formed by frontal bone.

Measure: ½ Anguli.

Type of Marma: Vaikalyakara Marma.

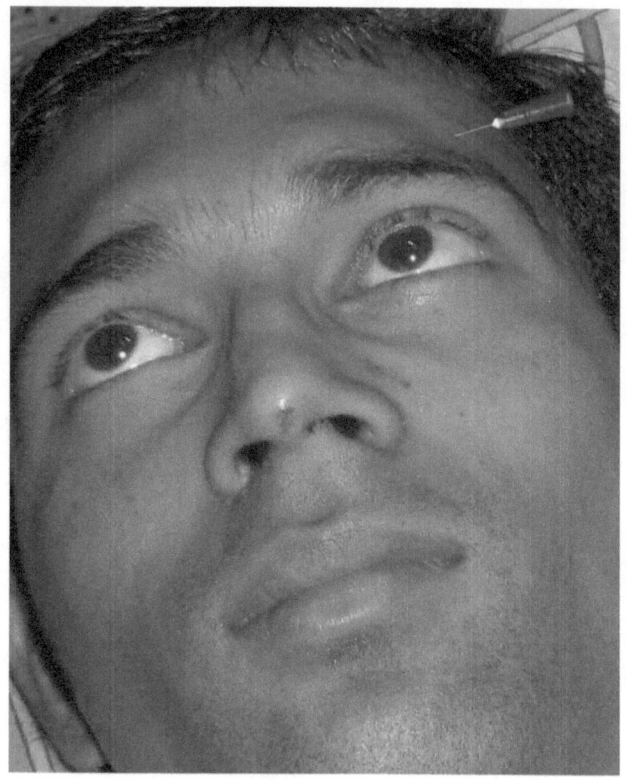

Signs of injury:
Injury may lead to disfigure the face.

In diseases of

- **Jivha** (Tongue)
- **Danta** (Teeth)
  Venipuncture should be done at
  the base of tongue by raising the
  tongue while keeping it constant

at a point by the help of upper incisors from inside.

In diseases of
- **Talu** (Palate)
- **Dantamula**
- **Kantha** (Throat)

In the diseases of palate, the puncture should be carried out on the palate by opening the mouth wide.

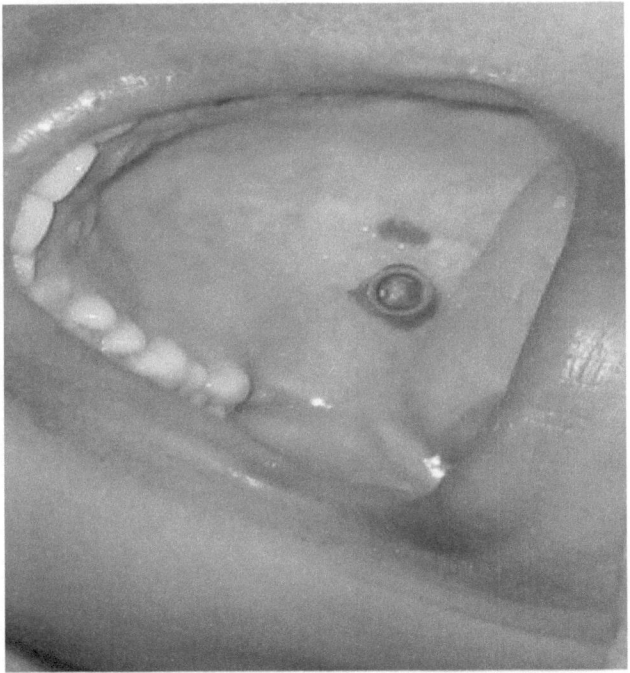

It should always be kept in mind that needle should never go inside the mouth otherwise it may lead to severe injury.

In the diseases of ears, puncture should be done above & around ears.

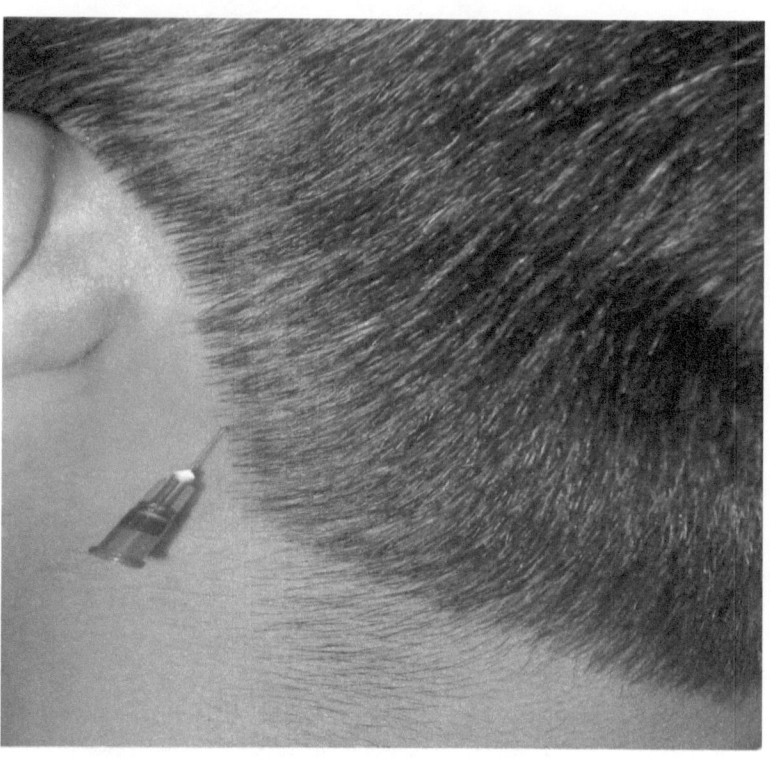

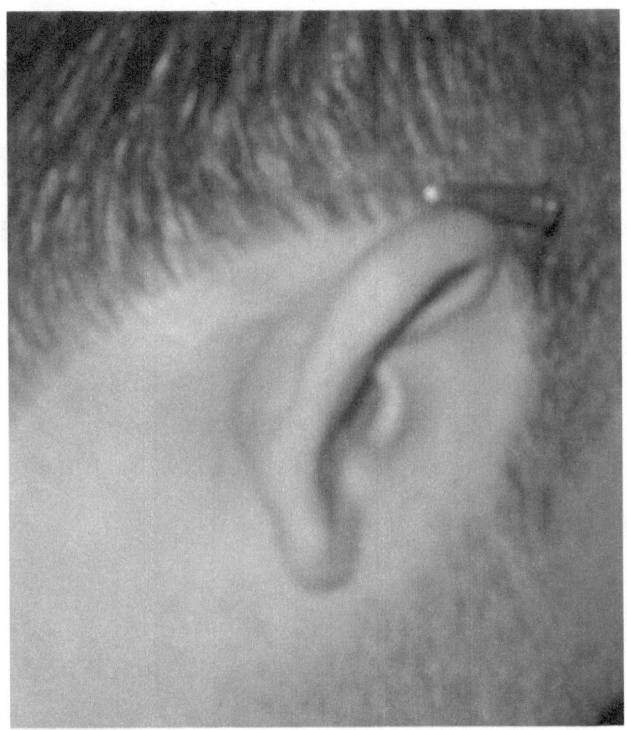

In case of

- **Anosmia**
- **Nasal disorders**

Vein should be punctured at the tip of nose.

In diseases like

- **Timira** (Black spots floating in front of eyes)
- **Akshi-Paka** (Suppuration of eyes)
- **Other eye diseases**
- **Siroroga** (diseases of head)
- **Adhimantha** (Glaucoma)

Viddha Karma points:
1) On the lateral side of nose.
2) Lateral end of eyebrow.
3) On the forehead above the eyebrow.

Anatomy

One finger above the midpoint of the eyebrow on the forehead, Viddha Karma should be done.

Anatomy for Phana Marma:

Location:
Nasa-Samipasthe - depression of the infra-orbital foramen.

In the level of saddle of the nose.

Measure: ½ Anguli.

Type of Marma:
Vaikalyakara Marma.

Signs of injury:
Injury may produce deformities of the nose & face.
Nerve may cause loss of smell sensation.

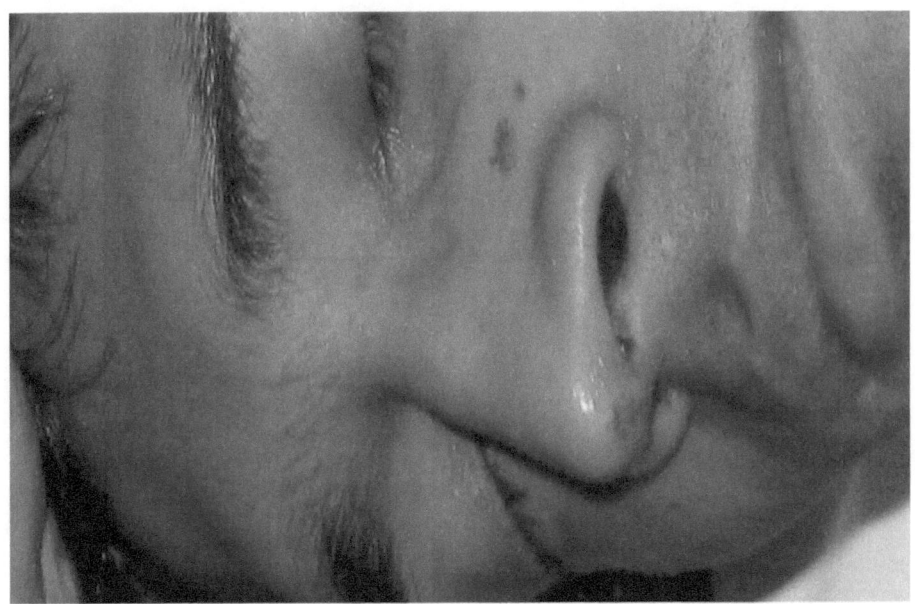

Nasa-Samipasthe

Anatomy of Apanga Marma:

Location: Depression of the lateral end of eyebrow (the outer corner of the eye).

Measure: ½ Anguli.

Type of Marma:
Vaikalyakara Marma.

Signs if injured:
Injury may lead to blindness & deformities of the face.

Signs of proper Viddha (Su. Sha. 8/11)

Viddha Karma is said to be proper if the blood comes out in a proper torrent in proper amount, stops by its own.

CHAPTER 3

Role Of Agnikarma In The Management Of Pain.

Introduction

The word itself means therapy/procedures in which Agni (Agni = fire), is used for the treatment. As Agni is Ushna (hot) in nature, it acts better for the treatment of Vataja Shula (pain). This is because; Vata is having Shita (cold) nature, opposite to Ushana property of Agni.

In case of Vataja Shoola (pain), when the pain is located at Snayu, Sandhi & Asthi, Snehana, Upanaha, Agnikarma, Bandha (tight bandaging) & Mardana is the treatment regimen. This treatment should be continued till the patient gets relief. Su. Chi. 4/8

Agnikarma is considered as the best therapy in Ayurveda. The conditions / diseases not curable by drugs should be sent for surgery. In case where surgery is also not successful, patient should be referred to Kshara Karma. If Kshara is also not giving the desired result, then the final line of treatment described in the texts is Agnikarma.

Points for performing Agnikarma:

General point of consideration for Agnikarma is, it should be done on some specific points. If the point of Agnikarma is not known properly, then

1. The points of Vaydha karma can be used for the same. OR
2. Another concept regarding the points of Agnikarma – the site of maximum pain is the ideal site for Agnikarma.

So, even in cases where the points are not stated, practitioners can select the site for Agnikarma as per the conditions of the patients.

Instruments for performing Agnikarma (Su. Su. 12/4):

- Pippali,
- goat's faeces,
- cow's tooth,
- arrow,
- rod,
- *Jambavaustha* (an instrument made up of stone),
- honey,
- jaggery &
- Other metallic instruments.

Among them, for the diseases located in -

1) Skin: *Pippali,* goat's faeces, cow's tooth, arrows are applicable.
2) Muscles: *Jambavaustha* & other metallic instruments should be used.
3) Blood vessels, ligaments, joints & bones, use of honey, jaggery & fatty substances are recommended.

When to perform Agnikarma (Su. Su. 12/5):

It can be performed in all the seasons except autumn & summer. However, in emergency conditions it can be performed in all seasons.

In today's practice, it is now done in all the seasons.

Signs Of Skin Burning:

A) Signs of skin burning (Su. Su. 12/8):
  1. Appearance of sound
  2. foul smell
  3. constriction of skin
B) Signs of muscle burning (Su. Su. 12/8):
  1. Pigeon's colour
  2. Little swelling
  3. Pain with dry & constricted wound.
C) Signs of blood vessels & ligaments burning (Su. Su. 12/8):
  1. black & raised wound
  2. stop of discharge
D) Signs of joints & bones burning (Su. Su. 12/8):
  1. Roughness

2. Reddish colour
3. Rigid wound

Types of Agnikarma marks (Su. Su. 12/11):

1. Valaya (circle)

2. Bindu (pointed)

3. Vilekha (line)

4. Pratisaran (flat)

In case of –

- **Shiro-roga**
- **Adhimantha**

Agnikarma should be carried out in eyebrows, forehead & Sankha Pradesh.

In case of –

- **Katishoola** (Low backache, Lumbar pain)
- **Grudrashi** (Sciatica)
- **Lumbar Spondylitis**
- **Spondylolisthesis** (Spondylolisthesis is the forward displacement of a vertebra, especially the fifth lumbar vertebra, most commonly occurring after a break or

fracture. Backward displacement is referred to as retrolisthesis).etc

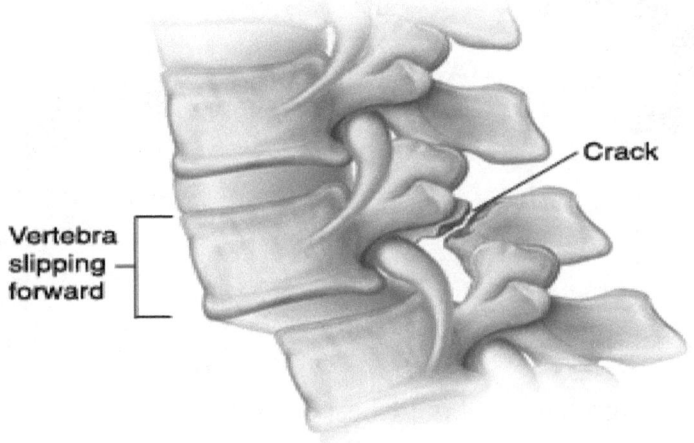

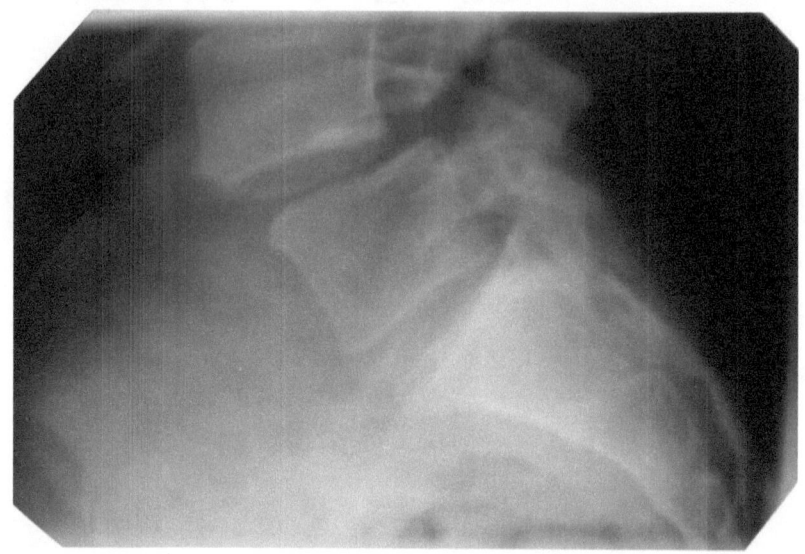

Agnikarma should be performed at the region 2 c.m. lateral to the level of L1-L2, L2-L3, L3-L4, L4-L5, L5-S1.

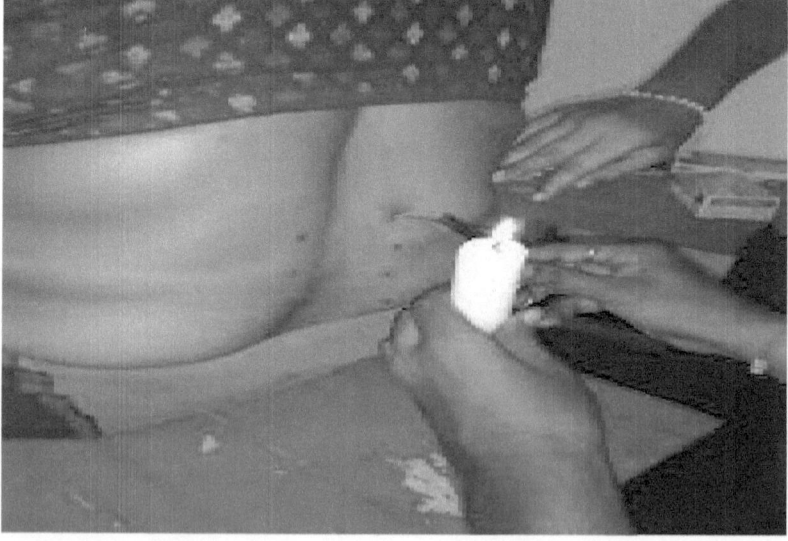

Anatomy:

Muscles & related nerves:

- Quadratus lumborum. It helps in Lateral flexion of vertebral column & is related to nerves T12, L1

- Psoas Major. It flexes thigh at hip joint & vertebral column & is related to nerve L2, L3, sometimes L1 or L4

In case of –

- **Janu-shoola** (knee joint pain) should be treated by performing Agnikarma at
  - ➢ Articular surface of Knee joint on medial & lateral sides.
  - ➢ Four fingers above & below the knee creases on the lateral sides.

Other indication is -   Osteo-Arthritis of Knee joint.

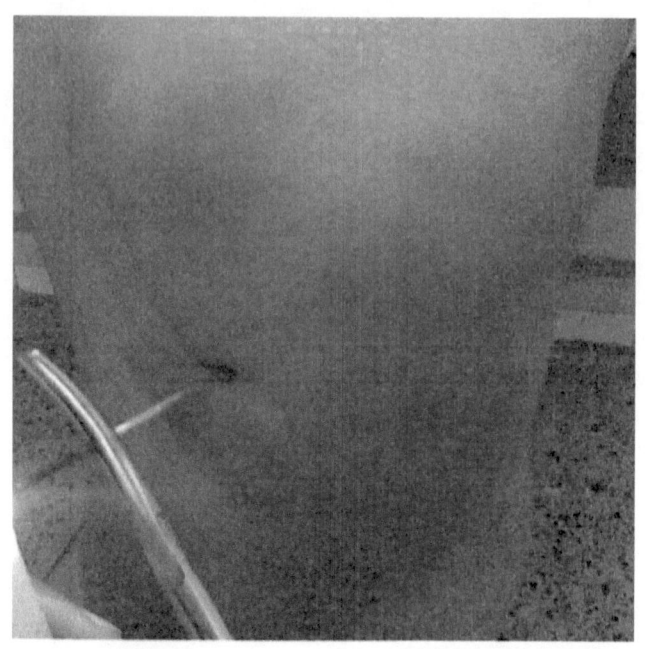

Anatomy:

1. The anterior cruciate ligament prevents the femur from sliding backward on the tibia (or the tibia sliding forward on the femur).
2. The posterior cruciate ligament prevents the femur from sliding forward on the tibia (or the tibia from sliding backward on the femur).
3. The medial and lateral collateral ligaments prevent the femur from sliding side to side.
4. Two C-shaped pieces of cartilage called the medial and lateral menisci act as shock absorbers between the femur and tibia.

Nerves: Femoral nerve

Lateral cutaneous nerve of thigh.

In cases of –

-   **Manya-shoola** (neck pain)
-   **Manya-sthambha** (neck stiffness)
-   **Vishvachi** (Brachial Neuritis)

Agnikarma should be performed at the lateral part of inter-vertebral disc spaces of cervical region.

Moreover, in case of Visvachi Agni-Karma should also be performed at two more sites; i.e. four fingers above & below the elbow joints.

Other indications:

Cervical Spondylolisthesis.

Cervical Spondylitis.

Etc.

Anatomy:

Muscles:

- **Apex:** Union of the sternocleidomastoid and the trapezius muscles at the superior nuchal line of the occipital bone

- **Anterior:** Posterior border of the sternocleidomastoid

- **Posterior:** Anterior border of the trapezius

Nerves: Cervical Nerves to trapezius

3rd & 4th occipital

Accessory nerve

In cases of –

- **Kurcha-shool** (Calcaneal Spur)

Agnikarma should be done on most tender point on calcaneum bone on the planter aspect.

Materials for Agnikarma: Mrittika Shalaka

Anatomy:

Calcaneal spurs are frequently associated with plantar fascitic, a painful inflammation of the fibrous band of connective tissue (plantar fascia) that runs along the bottom of the foot and connects the heel bone to the ball of the foot.

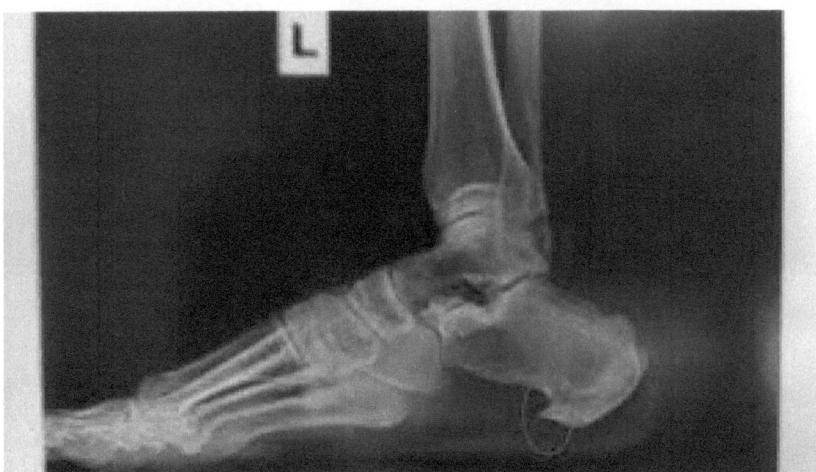

In cases of –

- **Manya-Granthi,**
- **Apachi** (Scrofula)

Agnikarma should be performed as Vilekha type (lining type) having 3 lines on the planter aspect of the wrist joint at a distance of 1 finger.

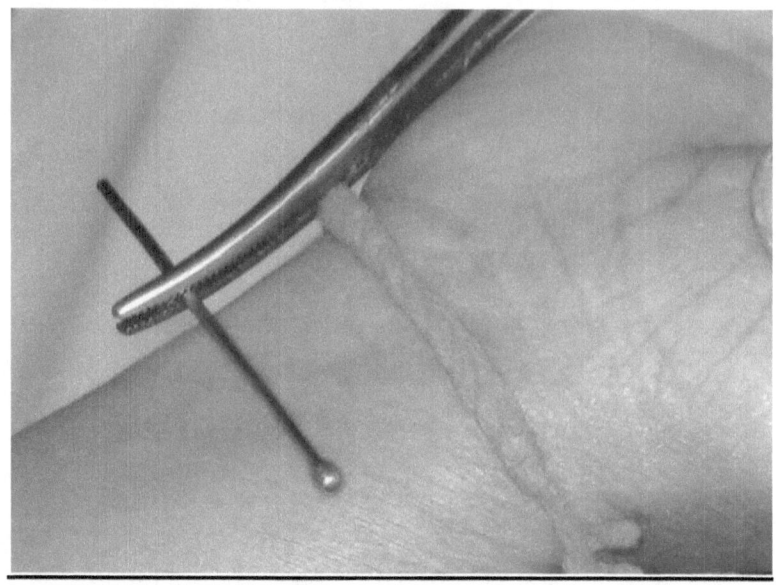

Anatomy:

Vascular: Medially - ulnar artery.

Radial artery & vein.

In cases of –

- **Yakrut-Vriddhi** (hepatomegaly)
- **Pliha-Vriddhi** (Spleenomegaly)
- **Kamla** (Jaundice)

Points of Agnikarma: Just at side of Anatomical snuff box of wrist in Yakrut Vriddhi &

jaundice. Agnikarma should be done on left side for spleenic enlargement.

In **Valmika** (Madura / Mathura foot), Naadi vrana should be excised & then Kshara treatment should be given or cauterization carried out.

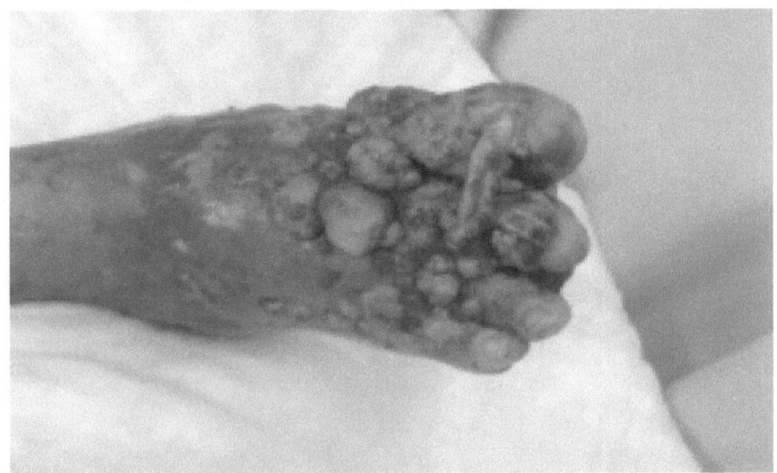

Iron Shalaka should be done red hot & then should be inserted inside the sinus. It gives some results.

Kadara (corn) should be treated by performing cauterising with hot oil or ghrita after the excision of Kadara. Or red hot iron Shalaka can be used to cauterise.

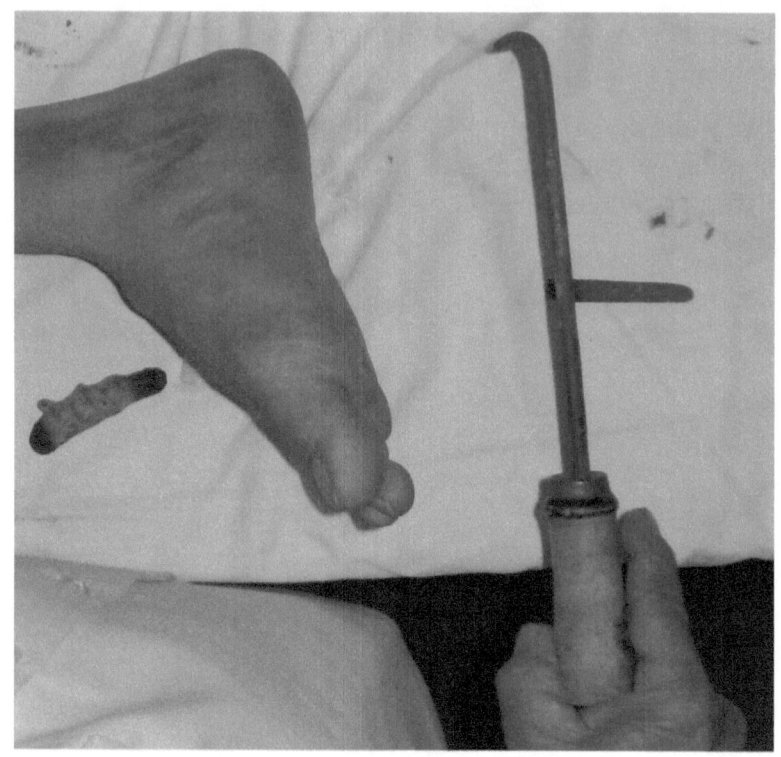

Chapter 4
Other Localised Treatments (Kati Basti, Griva Basti, Etc...).

So far, we have seen many therapeutic procedures mentioned in Ayurveda for the management of pain. But, there are some other specific local therapies also mentioned in Ayurveda which helps a lot in pain management. For example; pain of Cervical Spondylosis, Chronic pain in the neck region, sciatica, slipped and degenerative discs, back pain etc.

These local therapies are;

    a. Shiro Basti (Pooling medicated oil over head for sometime say half an hour)

    b. Griva Basti (Pooling medicated oil on back of neck for some time.)

    c. Prishta Basti (Pooling medicated oil on vertebral column for some time.)

    d. Kati Basti (Pooling medicated oil on lumbar region for some time.)

    e. Janu Basti (Pooling medicated oil on knee joint for some time.)

    f. Shirodhara (Pouring medicated oil/buttermilk/etc. on forehead.)

       Etc...

Though these are localised therapies, still helps a lot in coping local pains.

Note: Before applying the oils, one must be very clear about the Avasthas (Stages) of the diseases. It is a well known fact that Ayurveda considers "Aama" as the causative factor for almost all kinds of diseases. Aama is the toxic substances created in the body and later giving birth to different type of diseases. Aama has the properties similar to *Kapha,* but is toxic in nature. Oil or Ghee has properties of aggravating Kapha, and so Aama too. Hence in the presence of Aama, oils or ghee must not be used. If required, *Maha-Visagarvha tail or Vishagarbha Taila* can be used.

Indications for performing Kati Basti:

1. Slip disc
2. Backache
3. Sciatica
4. Intervertebral disc protrusion
5. Lumbar spondylosis
6. Lumbar spondylitis (Inflammation of lumber joints characterized by stiffness, tenderness and pain)
7. Degenerative disc disease

Procedure for performing *Kati Basti*;

Kati Basti is the procedure for relieving pain and other degenerative conditions of the body as mentioned earlier. It helps in relieving tension of the back muscles.

In this therapy the patient is told to lie on their stomach. After that a wall is created on the back at the point of deformity with dough of *Urid Daal*. The medicated oil is then warmed and poured inside the wall. When the oil is cool, it is then replaced again and again by luke warm oil. After that, steam is given. This process is continued for ½ hours on daily or weekly basis.

Materials required for Kati Basti

1. Black gram flour/powder = Approx. 150 grams (required every day) for making dough.
2. Medicated Oils = approx. 350 ml. This oil can be used 3-4 times in same individual.
3. Steel or plastic Ring = 2-3 inches in height and around 20 inch circumference.

Selection of medicated oils for the treatment

Sesame Oil for

- Acute/new backache with less severity.
- Back pain where only Vata is vitiated and only Vata pacifying properties are required.
- In case of back muscle stiffness.

Bala Oil or Balaswagandhadi oil

- Osteoporosis or degenerative conditions
- To give strength to the spinal vertebrae, muscles and ligaments.
- To eliminate vitiated Vata.

Mahanarayan Oil

- Degenerative conditions
- Mild discomfort and back debilities.
- Pain in back regions

Murivenna oil

- Excessive pain due to disk prolapsed
- Pain due to nerve compression

Nirgundi tail

- Low backache
- Muscle and spinal stiffness.

Kottamchukkadi Thailam

- Neuro muscular pains
- Sciatica
- Spondylosis

Precautions to be taken after Kati Basti

1. Minimum rest of 30 to 45 minutes after the procedure should be taken.
2. Avoid bending forward or backward after the Kati Basti treatment.
3. Lifting heavy weight after the procedure should be avoided.
4. Follow erect posture rule while sitting, walking, and standing.
5. Same oil used in one individual should not be used in other individuals. Otherwise it may pass on infection to others.

Duration & frequency of the procedure;

This treatment lasts for 45 minutes to 01 hour and it is good for any type of back pain and spinal disorders.

Similar treatment done at chest for curing heart diseases is called **'UROVASTI'** and same done on knee joints is called as **'JANUVASTI'. Same procedure done on** back of neck in cases of cervical spondylosis is known as 'GRIVABASTI'

Oils & Ghee used for localised Vasti

- Mahamash Tailam – One of the powerful Ayurveda oil widely used for many neurological conditions. It is also used for cervical spondylosis treatment.
- Kottamchukyadi oil
- Tila taila (sesame oil)
- Balashwagandhadi Tailam
- Ashwagandhabalalakshadi Tailam
- Sahacharadi tailam
- Ksheerabala tailam
- Murivenna oil
- Pinda tailam
- Tiktaka ghritam
- Mahanarayana tailam
- Guggulutiktaka ghritam
  Etc

These basic concepts have been provided in this book for the management of pain. These basic concepts can be easily practiced in today's clinical practices.

CHAPTER 5

REFERENCE BOOKS

1) Sushrut Samhita with Nibandhasangraha commentary of Sri Dalhanacharya, by Dr. Jadavji Trikamji, Chaukhambha orientalia varanasi, 2009

2) Sushrut Samhita – English Tranalation with Commentary of Shri Dalhanacharya, by Dr. Priyavat Sharma, Chaukhmba Sanskrit series, Varanasi, 2010

3) Viddha & Agnikarma Chikitsa, by Dr. R. B. Gogate, GFAM, AVP, Vaiyamitra Prakashana, Pune, Jan 2004.

4) Current Medical Diagnosis & Treatment 2013, by Michael W. Rabow, MD & Steven Z. Pantilat, MD, Chapter 5 on Palliative Care & Pain Management.

5) Internet websites, from Wikipedia, the free encyclopedia and other websites.

CHAPTER 6

KEYS FOR ABBREVIATIONS

Su.   = Sushruta Samhita

Su. Su = Sushruta Samhita Sutrasthana

Sa.   = Sharir Sthana

Ni.   = Nidana Sthana

Chi.   = Chikitsa Sthana

www.ingramcontent.com/pod-product-compliance
Lightning Source LLC
Chambersburg PA
CBHW021909170526
45157CB00005B/2022